'LEARN FROM THE PAST
HOPE FOR THE FUTURE'

Welcome to the Royal Armouries Museum.

All of us hope that you have a worthwhile and enjoyable visit.

The Royal Armouries is Britain's oldest museum. We hold in trust

for you one of the greatest and most comprehensive collections of

arms and armour in the world. This museum is not just about

battles, conflict and bloodshed, but about art, technology and

craftsmanship, about how people have lived and what they have

believed and held dear, about preserving life as well as taking it.

Our stories are stories that are as old as humankind and span the

whole world. We tell them through the display of our wonderful

collections; through the films, sound shows and computer

programmes that you will find in all the galleries; and especially

through the live demonstrations and dramatic 'interpretations' that

are such a special and unique feature of what we do.

The subject matter of this museum is a major part of human history;

and, we believe, an understanding of it, and of ourselves,

is of vital importance to our future. We hope by the time you leave

you will share our view. Thank you for visiting and please come

back soon. Every day is different in the Royal Armouries Museum.

Guy Wilson
MASTER OF THE ARMOURIES

CONTENTS

HISTORY OF THE ROYAL ARMOURIES

'The armoury where all manner of arms are kept in readiness'
King Christian IV of Denmark

The Royal Armouries is Britain's oldest national museum, and one of the oldest museums in the world.

It began life as the main royal and national arsenal housed in the Tower of London. Indeed the Royal Armouries has occupied buildings within the Tower for making and storing arms, armour and military equipment for as long as the Tower itself has been in existence.

Although distinguished foreign visitors had been allowed to visit the Tower to inspect the Royal Armouries from the 15th century at least, at first they did so in the way a visiting statesman today might be taken to a military

Sir Henry Lee, Master of the Armouries, 1578–1610. I.379

base in order to impress him with the power of the country. In the reign of Queen Elizabeth I less exalted foreign and domestic visitors were allowed to view the collections, which then consisted almost entirely of relatively recent arms and armour from the arsenal of King Henry VIII. To make room for the modern equipment required by a great Renaissance monarch Henry had cleared the Tower stores of the collections of his medieval predecessors.

The Tower and its Armouries were not regularly opened to the paying public until King Charles II returned from exile in 1660. Visitors then came to see not only the Crown Jewels but also the 'Line of Kings', an exhibition of some of the grander armours, mounted on horses made by such sculptors as Grinling Gibbons, and representing the 'good' Kings of England, and the 'Spanish Armoury', containing weapons

The Horse Armoury in the Tower of London by Rowlandson, about 1800, showing the Line of Kings. I.327

and instruments of torture said to have been taken from the 'Invincible Armada' of 1588. The Royal Armouries had become, in effect, what it has remained ever since, the national museum of arms and armour.

During the great age of Empire-building which followed, the collections grew steadily. Until its abolition in 1855, the Board of Ordnance, with its headquarters in the Tower, designed and tested prototypes, and organised the production of huge quantities of regulation arms of many sorts for the British armed forces. Considerable quantities of this material remain in the collections today, and some can be seen on the walls of the Hall of Steel.

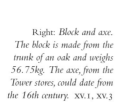

Right: *Block and axe. The block is made from the trunk of an oak and weighs 56.75kg. The axe, from the Tower stores, could date from the 16th century.* XV.1, XV.3

Also, throughout this period trophy weapons of all sorts continued to be sent to the Tower and displayed as proof of Britain's continuing military successes.

Early in the 19th century the nature and purpose of the museum began to change radically. Displays were gradually altered from exhibitions of curiosities to historically 'accurate' and logically organised displays designed to improve the visitor by illuminating the past. As part of this change items began to be added to the collection in new ways, by gift and purchase, and this increased rate of acquisition has continued to this day.

In this way the collection has developed enormously, the 'old Tower' material being joined in the last 150 years by the worldwide material which now makes the Royal Armouries one of the greatest collections of its type in the world.

Artist's impression of the Tower of London as it was in about 1890.

As the museum's collections continued to expand the Tower became too small to house it all properly. In 1988 the Royal Armouries took a lease on Fort Nelson, a large 19th-century artillery fort near Portsmouth. This is now open to the public and displays the collection of artillery.

In 1990, after two years of preliminary research and deliberation, the decision was taken to establish a new Royal Armouries in the north of England in which to house the bulk of the collection of worldwide arms and armour, thus allowing the Royal Armouries in the Tower to concentrate upon the display and interpretation of those parts of the collection which directly relate to the Tower of London. The concept of the Royal Armouries in Leeds had been born.

The new museum has been developed specifically to show the collections of the Royal Armouries in the best possible way. We began with the question 'How do we want to display our collections?', and the answer to that has dictated the sort of building which has been designed and built.

The Royal Armouries Museum has been built for the 21st century using the best of traditional museum design and it has been developed quite consciously to show its collections in relation to the real world in which we live. The displays seek to make the historical stories relevant by bringing them up to the present day. The building has, quite literally, been designed around the collections of the museum.

Above: *The Armouries as an arsenal: the former Cannon Room in the basement of the White Tower.*

Left: *The North Mortar Battery at Fort Nelson, containing three British 13-inch mortars of the 1860s.*

The displays are intended to entertain and stimulate a desire to learn, and our intention has been to create a multi-layered experience to cater for the many different interests and interest levels of our visitors.

The use of violence by humankind for supremacy or survival, or its sublimation into sport or play always has been, and probably always will be, one of the main forces for historical change. This is the underlying theme of the new Royal Armouries. It is a fascinating and often disturbing story of great importance to us and our children.

'Horrible war, amazing medley of the glorious and squalid, the pitiful and sublime, if modern men of light and leading saw your face closer, simple folk would see it hardly ever.'

Winston Spencer Churchill

War is one of those activities which most distinguishes the human species from other animals. Throughout the ages humankind has devoted much of its energy, ingenuity and power of social organisation to the systematic destruction of its own kind for reasons which may become incomprehensible with the passage of years but which were passionately felt at the time. As a result this century has seen humankind come within a button-push of utter destruction.

The face of battle seems to have changed enormously from the days of bloody hand-to-hand combat with which the gallery begins to the modern, sophisticated, technology-based warfare witnessed recently in the Gulf War. This has largely been due to the enormous developments in Western science, technology and industry since the Renaissance five centuries ago. It is easy to be seduced by the clinical precision and distant impersonality of modern 'high-tech' war and to forget that the majority of wars today are bloody, close, hand-to-hand affairs fought with relatively 'low-tech' weapons. For the inhabitants of the areas in which they are fought such warfare causes the world to be turned upside down just as it has from the beginning of human history. These wars have not really changed, even though the weapons have, and the act of warfare remains ever the same and essentially de-humanising. Yet, and here lies one of the fascinations of the study of war, in this apparently barren environment men and women through the ages have often displayed the most amazing courage and fortitude and the most sublime humanity and self-sacrifice. This gallery is their story.

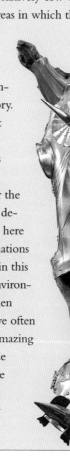

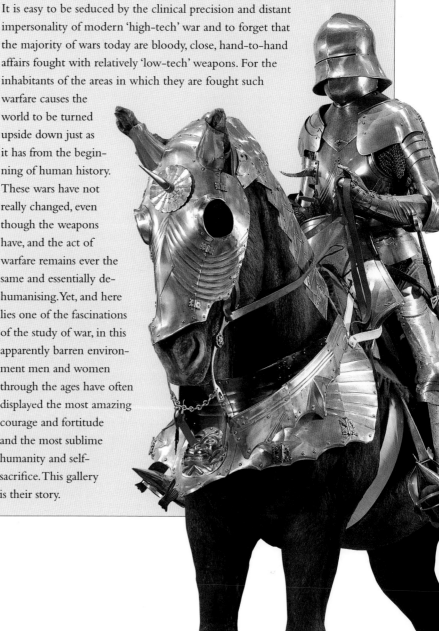

Above right: *View of the War gallery showing the Men at Arms display, representing armour and equipment of the 15th, 16th and 17th centuries.*

Right: *Armour for man and horse in the German 'Gothic' style, late 15th century. The horse armour was made for Waldemar VI, Duke of Anhalt-Zerbst, 1450–1508.* II.3, III.69, 70, 1216, 1300, VI.379

ANCIENT AND MEDIEVAL WARFARE

The entrance to the gallery is through 'Pomp and Ceremony', a display mainly of images evoking the romance of war depicted in art. This contrasts with the war cinema, in which an introductory film presents the subject of the gallery in a more realistic light, and takes the viewer back in time to the ancient world where the story of the gallery begins.

The museum has only a small collection of ancient arms and armour, which is displayed in the early war section together with a large number of images arranged to give a sense of the nature of ancient warfare and the arms and armour used in it.

The medieval section deals with the period in which the man-at-arms dominated the battlefield, but was compelled by developments in the technology of infantry weapons and tactics to wear heavy armour of steel plate. The first section of this display covers the period of the Hundred Years War, while opposite are displays of arms and armour of the period of the Wars of the Roses.

The developments of warfare in the early 16th century are represented by the battle of Pavia, 1525, a battle of significance in military history as one of the first occasions when mounted knights were defeated by infantry armed with firearms. The story of this battle is illustrated with a diorama of the battle and a near contemporary painting, as well as computer games simulating the tactics of the battle.

One of three surviving great helms of a characteristic English type of about 1360. Soon after this date the helm ceased to be used on the battlefield but continued in the tournament. IV.600

Probably the finest surviving example of the basinet, this one was made in Milan in the late 14th century. IV.470, bequeathed by Sir Archibald Lyle in memory of his two sons who were killed during World War II.

Above left: Model of an Italian export armour of the middle of the 15th century, based on the effigy of Richard Beauchamp in St Mary's Church, Warwick, cast about 1453. II.194

Left: A painting of the battle of Pavia, 1525, painted quite soon after the battle, which illustrates the arms and armour of the period and the events of the battle. I.142

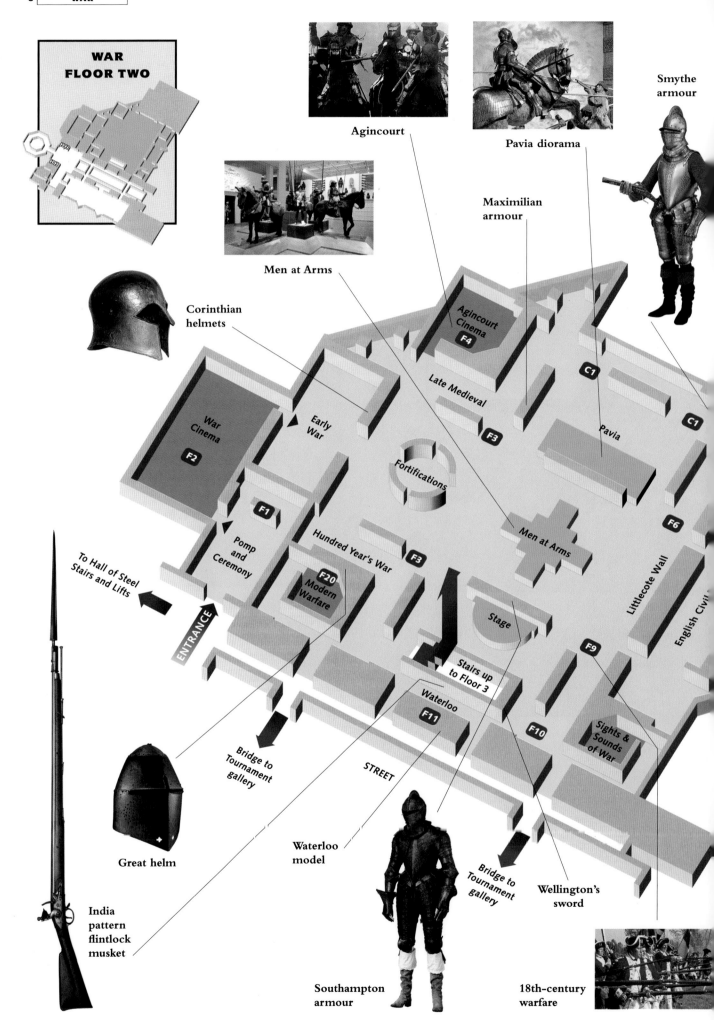

WAR
FLOOR TWO

Agincourt

Pavia diorama

Smythe
armour

Maximilian
armour

Men at Arms

Corinthian
helmets

Agincourt
Cinema
F4

Late Medieval

C1

C1

War
Cinema
F2

Early
War

F3

Pavia

Fortifications

Men at Arms

F6

F1

Pomp
and
Ceremony

Hundred Year's War

F3

Littlecote Wall

English Civil

To Hall of Steel
Stairs and Lifts

F20
Modern
Warfare

Stage

F9

ENTRANCE

Stairs up
to Floor 3

Waterloo

F11

F10

Sights &
Sounds
of War

Bridge to
Tournament
gallery

STREET

Bridge to
Tournament
gallery

Wellington's
sword

Great helm

Waterloo
model

India
pattern
flintlock
musket

Southampton
armour

18th-century
warfare

Gatling gun

Incident at
Rorke's Drift

WAR
FLOOR THREE

Dragunov sniping rifle

Calthorpe
sword

Zulu
War

F17

F16

C3

Wargames
Table

Prototype
Maxim gun

Early machine
guns

Crimean War

C2

F15

F14

American
Civil War

F13

To Lifts

Stairs down
to Floor 2

C4

F12

World War I

STREET

World
War II **F19**

Bridge to
Tournament
gallery

ENGLISH CIVIL WAR

The Royal Armouries includes one of the finest
collections of arms and armour of the mid 17th
century in the world. In addition to the arsenal
collection, preserved at the Tower of London
since the time of its use, the museum acquired in
1985 the armoury from Littlecote House, the last
major Civil War armoury then remaining in
private hands. This collection is important for
its groups of buff coats, baldricks and carbine

slings, and for its
muskets, carbines
and pistols. Some
of the pieces from
Littlecote are
displayed on a
reconstruction of
part of the Great
Hall at Littlecote
House, where the
armoury was dis-
played from at
least the late
17th century.

WWI Machine gunner

COMPUTERS

C1 Pavia

**C2 American Civil
War – Sharpsburg**

C3 Zulu – Isandlwana

C4 The Great War

FILMS

**F1 Pollock Theatre –
Pomp & Ceremony**

**F2 Introduction to
the War gallery**

F3 Bows & crossbows

F4 Agincourt

F5 Armada

**F6 How a man
schal be armyd**

F7 Civil War firearms

F8 Marston Moor

**F9 18th-century
warfare**

F10 Culloden

F11 Waterloo

**F12 The Age of
Invention**

**F13 The American
Civil War**

F14 Crimean War

**F15 The Cavalry
Sword**

F16 Mass production

F17 Zulu

F18 World War I

F19 Word War II

F20 Modern Warfare

THE 17TH AND 18TH CENTURIES

Right: *Equipment for an English harquebusier of about 1650, comprising a buff coat, harquebusier's pot, 'mortuary' sword and baldrick, carbine and carbine sling, all from the armoury of Littlecote House.* III.1942, IV.887, IX.2789, 3356, XII.5480; XIII.303, 313

Far right: *India pattern flintlock musket of the period of the Napoleonic Wars. This was the standard British infantry weapon for nearly a century.* X.90; XII.3507

Right: *Armour of the military writer and commander Sir John Smythe made in the royal workshops at Greenwich, about 1585, shown with a contemporary wheellock pistol.* II.84 III.1470–1; XII.716

Below right: *A scene of British troops advancing with fixed bayonets, from the 18th-century warfare film.*

By the later years of the 16th century European armies were using more co-ordinated groups of specialist troops such as musketeers and pikemen. Armour and weapons of officers and soldiers of the period are shown in the context of the defeat of the Spanish Armada of 1588, and a short film explains the development of naval gunnery by that time.

The wall adjacent to this display is decorated with some of the equipment from Littlecote House, in Wiltshire, the last complete English Civil War armoury surviving in England, acquired by the Royal Armouries in 1985 to prevent its dispersal at auction. Beyond this wall is a collection of early 17th-century equipment for officers, and that used by cavalry and infantry during the English Civil Wars.

By the end of the 17th century fixing a stout knife into the muzzle of a musket had produced the first bayonet, and gradually this development replaced the pike, which had been such a feature of the battlefield for over 200 years. Infantry armed with a musket which could be used as a pike once it had been fired caused changes in tactics: the 18th century saw the development of 'linear' warfare, the deployment of bodies of infantry in tight formation firing in volleys and then charging with fixed bayonets. Tactics like these were employed during the American War of Independence and most dramatically in Britain's wars with France at the end of the century.

The weapons of that period are displayed near a unique mid 19th-century model of the battle of Waterloo, showing the distribution of troops at about 2 pm on the day of the battle, 18 June 1815.

THE 19TH AND 20TH CENTURIES

The upstairs floor follows the extraordinarily rapid advance in weapons technology from the early 19th century to the end of the Second World War. The earliest cases look at the rapid change from flintlock musket to breechloading rifle that occurred in a period of only forty years, between 1830 and 1870, and the adoption of the American system of mass production.

The displays cover the major conflicts of the late 19th and 20th centuries. There are also special displays illustrating the advances in smallarms design which examine the development of the bayonet, the cavalry sword, cartridges and the art of sniping.

The sword presented to Major Somerset Calthorpe of the 8th Hussars, aide-de-camp to Lord Raglan, on his safe return from the Crimean War in 1855. IX.2603

Cases showing the earliest repeating magazine rifles of the 1870s are followed by a display tracing the later development of the modern magazine-fed, centrefire rifle. There are also some examples of the quirkier forms of repeating firearms that were not always as successful as their inventors had hoped. The birth of automatic weapons is shown with Hiram Maxim's first machine-gun, along with a selection of later automatic weapons, taking the visitor into the World War I area, with its impressive display of trench weapons and tableau of a Vickers machine-gunner.

Next to this display are cases that trace the later development of the bayonet and some of the firearms of the inter-war period. The final cases on this level show some of the huge range of weapons used during World War II.

The story closes downstairs where cases showing recent developments in postwar and modern weapons and equipment can be seen in the area around the foot of the staircase.

Above: *A Colt Model 1851 Navy revovler of 1856. A special order London-made example with foliate engraved barrel, cylinder and frame, which has a fitted oak case.* XII.1434

Left: *A Dragunov SVD sniping rifle of about 1975. A semi-automatic rifle based on the famous AK-47 design, it was adopted in the mid-1960s as the standard sniping rifle by most Eastern Bloc military powers.* XXIV.8853

Far left: *A .65-inch Gatling gun on a light field carriage designed to be pulled by infantry. Made by Sir W G Armstrong & Co., Newcastle-on-Tyne, for British service, 1873.* XII.1804

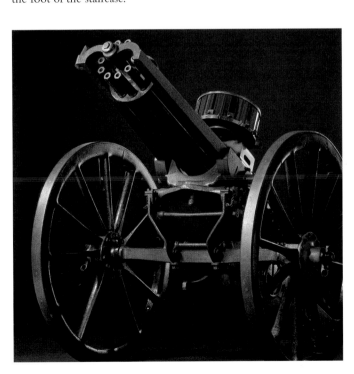

TOURNAMENT

'Trulie, this action was mervailouslie magnificent, & appeared a sight exceeding gorgeous.'

From a description of a torchlight tourney held in Whitehall in 1572 in the presence of Queen Elizabeth I

The story of the tournament is the story of the development of a form of early medieval practice for war, which was generally frowned on by monarchs as a danger to public order, into perhaps the greatest of the Renaissance courtly entertainments used by monarchs to demonstrate their power, wealth and chivalric ideals. From uncontrolled beginnings it developed into a highly regulated and sophisticated sport with as many different varieties as football. From the 15th century onwards special weapons and very specialised armours were required for the various types of tournament event, and the Royal Armouries contains a fine collection of these. They are exhibited and interpreted in this gallery which charts the development of the three major types of tournament fighting – the tourney, the joust and foot combat.

Accidents, such as the death of the King of France after a joust in 1559, reduced the popularity of the tournament but it was really the rise and domination of firearms on the battlefield in the 17th century that finally saw the demise of the tournament.

Above: *Reconstruction of a tent housing the tonlet armour of Henry VIII, which was almost certainly made in the royal workshop at Greenwich for the King to wear in the foot combat at the Field of Cloth of Gold tournament of 1520.* II.7

Right: *The Lion armour, an embossed and damascened armour made in Italy or France, about 1545–50. A detail is also shown.* II.89

THE TOURNAMENT

The gallery is set out as a tournament field with two tents, a tree of honour and a foot combat arena. At one end of the gallery there are video screens which explain the three forms of tournament combat: the tourney, the joust and the foot combat.

The tourney was fought as a mock battle. It became a team event fought first with lances and then with swords; the contestants wore battle armour with many extra reinforcing pieces.

The joust was a contest between one mounted individual and another. There were two forms of joust: the joust of war with the aim of unhorsing and the joust of peace with the aim of shattering lances.

Demonstration in the foot combat arena in the Tournament gallery.

In the red tent is an astonishing armour used for the joust of war. It was made for a tournament held by Maximilian I. This armour has a special effect. If the armour was struck over the brow or on the lance-guard, then pieces of the armour flew off giving the impression of breaking metal.

The joust of peace was fought with armour including a large helm bolted down to the breastplate and backplate. A large slot at the front of the helm allowed the jouster to take aim by leaning forward. He then straightened up to protect his eyes from broken lance pieces. This meant that at the point of impact he could see nothing and could only feel the hit.

The foot combat was a contest fought on foot with a variety of weapons. Near the foot combat arena is a blued and gilt armour made as part of an eagerly anticipated Christmas present for the Elector of Saxony in 1591. Sadly, he died in September that year. It has no leg defences as it was designed to be used over a barrier. No expense was spared for these events, as can be seen from the two small armours made for a pair of brothers aged about 8 and 10, even though they would grow out of them very quickly.

Above: *'Frog-mouthed' helm for the joust of peace. The projecting lip protected the face from flying wooden splinters.* IV.411

Below: *Foot combat armour of the Elector Christian I of Saxony, made by Anton Peffenhauser of Augsburg, 1591. There are no leg defences as this form of combat was fought over a barrier.* II.186

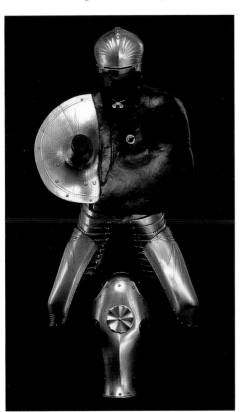

Left: *Joust of war armour from the court of the Holy Roman Emperor Maximilian I. The 'blind' shaffron helped to prevent the horse from shying at the moment of impact.* II.167

THE TOURNAMENT

There is a special gallery display showing the greatest tournament ever held between Henry VIII and Francis I of France called the Field of Cloth of Gold (named after the tents which were made of gold cloth). Here are two of the armours of the young, athletic Henry VIII and a film showing the King being dressed in one of the armours. These two armours of Henry VIII are among the earliest products of the royal workshop at Greenwich which was founded by him.

The armour of Robert Dudley, Earl of Leicester, stands in a setting reminiscent of a Tudor Palace. He was the favourite of Queen Elizabeth I and was allowed the privilege of ordering an armour from the royal workshops. The bear motif on the shaffron (horse's head-defence) is the only example of true embossed decoration from the Greenwich workshop. The museum commissioned a textile covering celebrating the bear which can be seen lining the walls of the Gallery.

In the corner of the gallery, stairs lead up to the next floor. This upper level, decorated with graphic walls, allows a fine view of interpretations taking place in the arena. There are computer interactives, a display of fakes and the story of the disastrous Eglinton Tournament of 1839.

The Lion armour stands in a Renaissance setting. This is the finest decorated armour in the Museum. It is covered in astonishingly intricate gold work and embossed all over with snarling lions' heads. This glorious armour could be mounted in different ways for various cavalry and infantry uses at tournament festivities and court events.

Above: The tonlet armour of Henry VIII for the Field of Cloth of Gold tournament in 1520. This is one of two foot combat armours made for the tournament. II.7, IX.633

Right: *Foot combat armour of Henry VIII. This armour was in production for the King to wear at the Field of Cloth of Gold, but was never completed.* II.6, VII.1510

The 'Burgundian' Bard was a gift from the Holy Roman Emperor Maximilian I to King Henry VIII, about 1510. This horse armour is decorated with devices from the Order of the Golden Fleece. VI.6–12

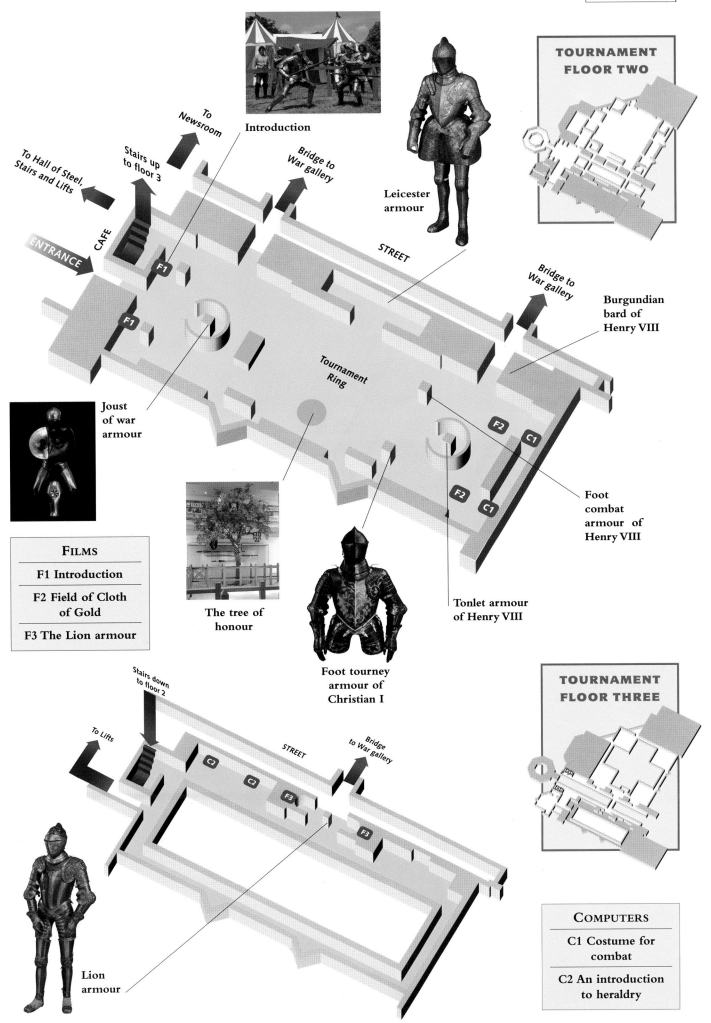

TOURNAMENT FLOOR TWO

Introduction

To Newsroom

Stairs up to floor 3

To Hall of Steel, Stairs and Lifts

CAFE

ENTRANCE

Bridge to War gallery

STREET

Leicester armour

F1

F1

Joust of war armour

Tournament Ring

Bridge to War gallery

Burgundian bard of Henry VIII

F2

C1

F2

C1

Foot combat armour of Henry VIII

The tree of honour

Tonlet armour of Henry VIII

Foot tourney armour of Christian I

FILMS

F1 Introduction
F2 Field of Cloth of Gold
F3 The Lion armour

Stairs down to floor 2

To Lifts

STREET

Bridge to War gallery

C2

C2

F3

F3

Lion armour

TOURNAMENT FLOOR THREE

COMPUTERS

C1 Costume for combat
C2 An introduction to heraldry

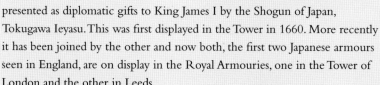

'Military action is important to the nation – it is the ground of death and life, the path of survival and destruction, so it is important to examine it.'

Sunzi The Art of War about 4th century bc.

Arms and armour have been made and used around the world for war, sport, military practice and self-defence. This gallery concentrates upon the great civilisations of Asia, and its purpose is to show how arms and armour can provide a key to understanding Asian history, a subject which until recently in Europe has been largely ignored.

The Royal Armouries acquired its first Asian piece in the 17th century – one of two armours presented as diplomatic gifts to King James I by the Shogun of Japan, Tokugawa Ieyasu. This was first displayed in the Tower in 1660. More recently it has been joined by the other and now both, the first two Japanese armours seen in England, are on display in the Royal Armouries, one in the Tower of London and the other in Leeds.

The Royal Armouries oriental collection was acquired in two main phases. Although a few pieces, military trophies and diplomatic gifts, had arrived earlier, it was not until the middle of the 19th century that Asian material was actively collected. By 1870 a very substantial collection had been built up, and was displayed in one of the main galleries in the White Tower. A second phase of collecting started in the 1960s and has continued until the present day.

The cultures of Asia are far more diverse than those of Europe, and the gallery is divided into a number of distinct zones – central Asia, Islam, the Indian sub-continent, China, Japan, and south east Asia. But there is one military theme that unites all these diverse cultures, and has dominated the way in which war was waged in most of them until the 19th century: the use of the mounted archer.

Above right: The line of equestrian figures running down the centre of the Oriental gallery. The nearest one is a Mughal Indian armour of the late 16th century, the next a composite Turkish armour of the late 15th century.

Right: The Mughal Indian elephant armour of about 1600.
XXVIA.102

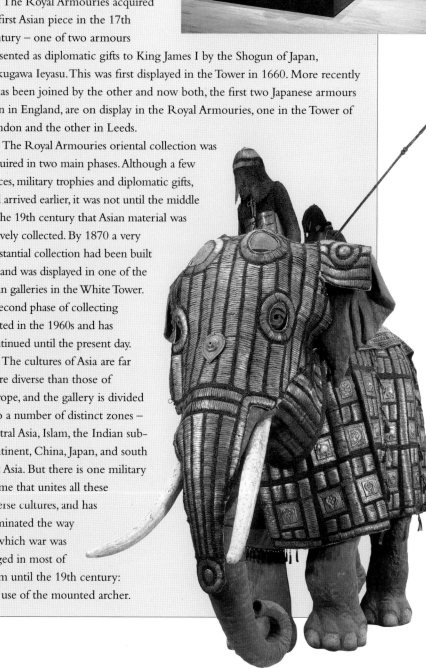

CENTRAL ASIA, ISLAM AND INDIA

The gallery is introduced by a small section on medieval central Asia. Many cultures introduced the style of warfare of the steppe nomads, based around the use of the horse archer, to the surrounding civilisations of west, south and east Asia. The most famous of these steppe armies were those of the Mongols, who under Genghis Khan built a massive world empire in the 13th century. Scarcely any Asian arms and armour survive from that time, but later material gives an insight into the equipment of the conquering armies, and the display includes a reconstruction, from original pieces, of a Mongol heavy cavalryman of this period. There is also a rare example of a Mongolian helmet of the conquest period.

Central Asian lamellar armour and other cavalry equipment, representing a Mongol heavy cavalryman of the 13th–15th century. XXVIA.122, 157, 276; XXVIB.141, 145; XXVIH.21–2, 38, XXVIS.298. *The horse armour is on loan from the Victoria and Albert Museum.*

The influence of the Mongol armies fell most heavily on the Islamic world, itself the product of an amazingly rapid period of empire building back in the 7th and 8th century. The display of Islamic arms and armour includes some important medieval armour, including a remarkably fine example of the 'turban' helmet, decorated with Arabic verses, complete with its mail aventail, and an important early Turkish sabre or *kiliç*.

A quoit turban or dastar bungga, *Indian, Lahore, 18th century. These turbans were worn by the Akali Sikhs, and carried a variety of throwing quoits* (chakram), *garrotting wires and knives.* XXVIA.60

Through the arch on the right can be seen the arms and armour from India, which form the largest part of the Royal Armouries Asian collection. This section includes the largest armour in the whole collection, the only elephant armour in captivity. Probably made in one of the arsenals of the Mughal Empire in northern India in the late 16th or early 17th century, in its present state, with two of its mail and plate panels missing, it weighs 118 kg. Contemporary with this is a complete armour for man and horse of the Mughal period, displayed alongside its Turkish counterpart.

Left: *Indian quilted armour of Tipu Sultan of Mysore, acquired from the collection of the Duke of York in the early 19th century.* XXVIA.139

Right: *A Turkish sabre* (kiliç) *of the mid 16th century. Made by a craftsman in Istanbul, the hilt was originally inlaid with precious stones.* XXVIS.293

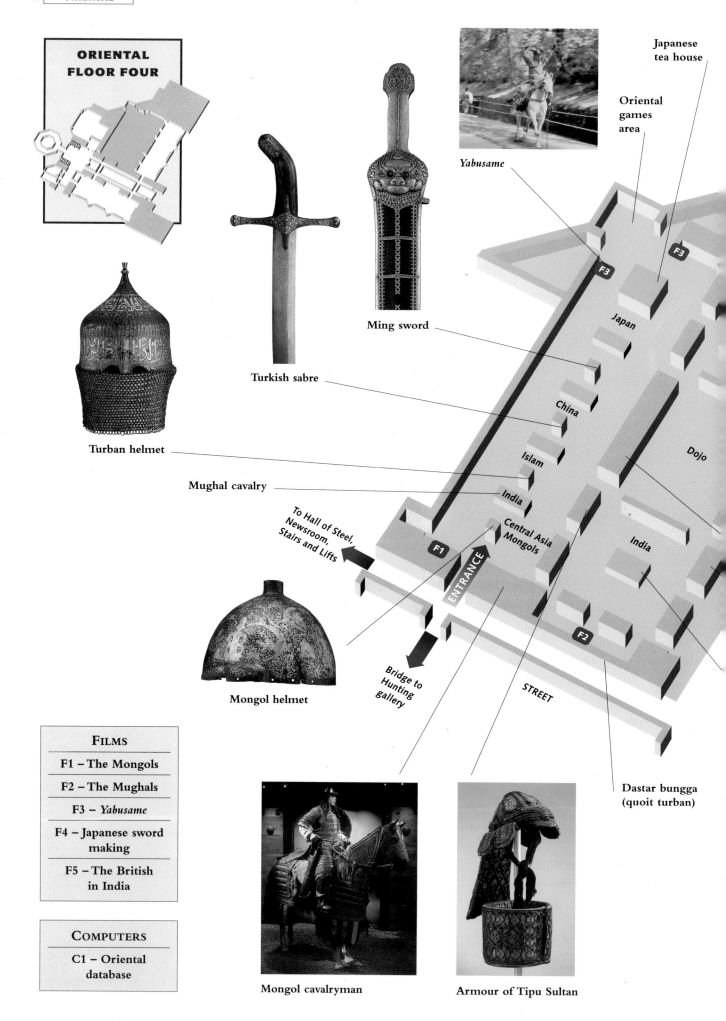

ORIENTAL FLOOR FOUR

Turkish sabre

Ming sword

Yabusame

Japanese tea house

Oriental games area

Turban helmet

Mughal cavalry

F3

F3

Japan

China

Islam

India

Dojo

India

To Hall of Steel, Newsroom, Stairs and Lifts

F1

ENTRANCE

Central Asia Mongols

F2

Bridge to Hunting gallery

STREET

Mongol helmet

Dastar bungga (quoit turban)

FILMS
F1 – The Mongols
F2 – The Mughals
F3 – *Yabusame*
F4 – Japanese sword making
F5 – The British in India

COMPUTERS
C1 – Oriental database

Mongol cavalryman

Armour of Tipu Sultan

JAPANESE ARMS AND ARMOUR

The arms and armour of Japan are very important in the collections of the Royal Armouries, and the very first Asian armour to enter the collection, by at least 1660, was Japanese, one of the armours presented to King James I by Tokugawa Ieyasu. Historical swords and armours are much more highly regarded in Japan than elsewhere because of the Shinto association of these objects with the spirits of the dead. The museum is also twinned with a Shinto shrine, the Nikko Toshogu Shrine, which is famous for its performances of *yabusame* (Japanese horse archery) events.

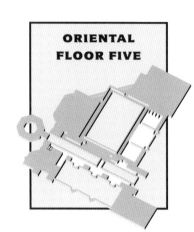

ORIENTAL FLOOR FIVE

Japanese gift armour

Algerian guns

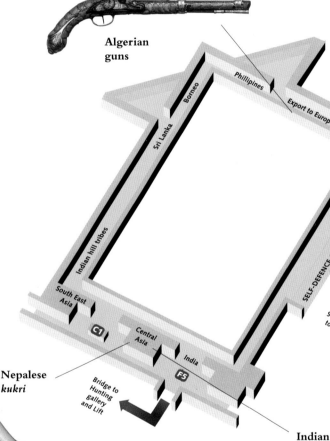

Borneo
Phillipines
Sri Lanka
Export to Europe
Indian hill tribes
SELF-DEFENCE GALLERY
South East Asia
Stairs down to Floor 4
C1
Central Asia
India
F5
Bridge to Hunting gallery and Lift

FENCE GALLERY
Stairs up to Floor 5

Chinese weapons

Pangolin armour

Nepalese *kukri*

Indian punch-dagger

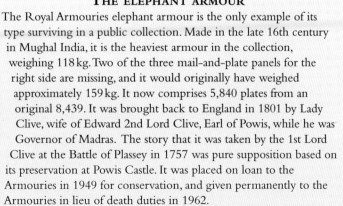

THE ELEPHANT ARMOUR

The Royal Armouries elephant armour is the only example of its type surviving in a public collection. Made in the late 16th century in Mughal India, it is the heaviest armour in the collection, weighing 118 kg. Two of the three mail-and-plate panels for the right side are missing, and it would originally have weighed approximately 159 kg. It now comprises 5,840 plates from an original 8,439. It was brought back to England in 1801 by Lady Clive, wife of Edward 2nd Lord Clive, Earl of Powis, while he was Governor of Madras. The story that it was taken by the 1st Lord Clive at the Battle of Plassey in 1757 was pure supposition based on its preservation at Powis Castle. It was placed on loan to the Armouries in 1949 for conservation, and given permanently to the Armouries in lieu of death duties in 1962.

CHINA AND JAPAN

View of the gallery, showing the Fukutokuan tea house and yabusame *figure lent by the Nikko Toshogu Shrine.*

The central area of the gallery is devoted to the arms and armour of China. Many military inventions came from China, including the crossbow and perhaps most important of all, gunpowder. The Chinese display includes a section on the history of early gunpowder weapons. Chinese arms and armour could also be great works of art. One sword displayed in this section, made for presentation to one of the great Buddhist monasteries of Tibet by a Chinese Emperor of the Ming dynasty in the early 15th century, is an example of the use of arms and armour in Asia as a vehicle for decorative art of the highest quality.

The arms and armour of Japan have perhaps the most prestigious history of any in the world. The museum has been twinned since 1991 with the Nikko Toshogu Shrine in Japan, the shrine built in 1623 as the resting place of the Shogun Tokugawa Ieyasu, who in 1613 presented to King James I the very first Japanese armours ever seen in England. The shrine is famous for its contests in traditional Japanese horse archery (*yabusame*), and a figure of a man and horse wearing the equipment used today in the *yabusame* festival at Nikko forms the centrepiece of this section of the gallery. Next to this is a reconstruction of a Japanese tea house, in which tea ceremonies can be performed. There are two films in the Japanese section of the gallery, one showing *yabusame* at Nikko, the other recording the forging of a Japanese sword by a traditional swordsmith.

Behind the Japanese and Chinese sections of the gallery is the Dojo (the Japanese name for a martial arts gymnasium), in which a variety of demonstrations of Asian and European martial arts, as well as other interpretations, are performed.

Chinese sword of about 1420, probably made for presentation by an early Ming emperor to one of the Buddhist monasteries of Tibet.
XXVIS.295

Above right: *One of the two Japanese presentation armours given by Tokugawa Ieyasu to King James I in 1613.* On loan from the Royal Collection.

Right: *A demonstration showing an armed samurai, using a modern replica of a medieval armour. Original armours are too fragile to use in this way.*

SOUTH AND SOUTH-EAST ASIA

To reach the upper floor of the Oriental gallery, either go out to the Street and take the lift up one floor, or go up the stairs in the Self-defence gallery.

The upper floor in the Oriental gallery continues the story of Indian arms and armour from downstairs. There are specific displays on European contacts with India and the Indian Mutiny, both of which reflect the museum's wealth of 19th-century Indian material. Rarer are the displays of south Indian weapons and those of Sri Lanka and the hill tribes of central India, which few museums are able to illustrate. The upper floor also continues the story of central Asian arms and armour, with sections on Nepal, Tibet and Bhutan.

Rarer still are the weapons of South-East Asia, to which two displays are devoted. One deals with the mainland, and includes an interesting collection of Burmese swords and exquisitely decorated staff weapons. The other covers the Malay peninsula and the islands of Indonesia, and includes a small collection of the characteristic sword of the region, the kris.

Finally there is a display devoted to finely decorated arms made in Asia for export to Europe, including a remarkable series of coral-mounted firearms made in Algeria for presentation to the royal courts of Europe, and a small group of smallswords made in India and Japan for export. This section leads naturally to the displays of finely decorated European swords in the Self-defence gallery next door.

Far left: *A Nepalese kukri with an ivory hilt carved in the form of a lion, made in the 19th century.* XXVID.30

Left: *An Indian punch-dagger* (katar) *with a double blade, confiscated in 1858 after the Indian Mutiny.* XXVID.70

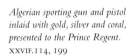

Algerian sporting gun and pistol inlaid with gold, silver and coral, presented to the Prince Regent. XXVIF.114, 199

A dagger from Darjeeling in Sikkim, the iron hilt gilt and silvered. Made in the 18th century. XXVID.36

SELF-DEFENCE

'The right of each to carry arms – and these the best and the sharpest – for his own protection is a right of nature indelible and irrepressible.'

James Paterson 1877

From the time of man's earliest existence, people have armed themselves against the threat of violence from others.

This gallery traces the story of the use of weapons in civilian life, and the ways in which society has attempted to control them and ensure that its citizens might be safe. Issues such as the development of police forces and the growth of prisons and the impact of mass-production in the manufacture of ever smaller, more compact weapons are fully illustrated from the Royal Armouries collections.

It explains, too, how for many centuries in Europe weapons were much more part of daily life than they seem today. Swords and daggers were openly carried by most classes in society, indeed they were often fashionable items in their own right and an essential part of everyday dress.

People are now increasingly aware that violence can threaten in the home from abuse by partners and parents, as well as from random attacks by strangers on the street.

Weapons have been kept and carried in ordinary life for other reasons besides self-defence and law enforcement: for instance, for sports such as target shooting and fencing; for duelling in defence of aristocratic honour; as fashion accessories; or to symbolise authority and status and commemorate courage and service.

By contrast, the upper part of the gallery displays some of the finest arms and armour ever made: a reminder of the tremendous skills of the craftsmen who made them.

Above right: Part of the Self-defence gallery explaining the historic connection between weapons and the law. In the foreground is the figure of a police dog trainer, who is wearing a special armour made of synthetic materials.

Right: The 'horned helmet' is all that survives of an armour made by Konrad Seusenhofer of Innsbruck in 1511–14 for presentation to King Henry VIII by the Holy Roman Emperor Maximilian I. IV.22

THE ARMED CIVILIAN

With no organised peacekeepers to uphold the law travellers, traders and pilgrims routinely armed themselves for protection. This gallery aims to show how weapons evolved with the advance of technology and changes in fashion. Simple daggers and swords developed into weapons both more efficient and more sophisticated, eventually becoming the elegant rapiers of the late 16th century. By the 1750s we can see how firearms, particularly pistols, had begun to replace the sword as a means of defence. Further displays also show some of the weapons that the traveller abroad would have carried, at a time when being armed was regarded as a matter of necessity.

A scene from one of the gallery's films.

As society became more regulated, so law and order became increasingly important and public servants were frequently armed. The central displays in the gallery look specifically at the weapons carried by police and prison officers, Customs and Excise officers and on Royal Mail coaches. A selection of some of the weapons and equipment now used by modern police forces can be seen.

Criminals have always carried weapons and the later cases look at some ingenious concealed and illegal weapons. Pistols and swords were also used for ritual and sporting purposes, such as duelling in the 18th century, or target shooting and fencing and this is reflected in the displays at the far end of the gallery. Two final cases trace the fascinating and important role of the firearm in helping to tame the American West, from the early pioneer days of the Kentucky flintlock musket through to the Colt semi-automatic pistol of the 20th century.

Far left: *A flintlock blunderbuss by Whitney of London, about 1775. A favourite weapon of the mail-coach guard, its bell-shaped muzzle caused its charge of lead shot to spread widely in a short distance.* XII.1042

Left: *An American powder horn of about 1770. Engraved with a map of the Mohawk and Hudson rivers, this horn is typical of the type used by the colonists during the Wars of Independence.* XIII.126

Below: *Small sword by Clare of London, about 1758. A very fine quality silver hilted gentleman's sword chiselled and pierced with motifs of foliage and trophies.* IX.798

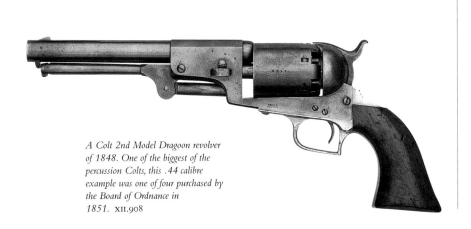

A Colt 2nd Model Dragoon revolver of 1848. One of the biggest of the percussion Colts, this .44 calibre example was one of four purchased by the Board of Ordnance in 1851. XII.908

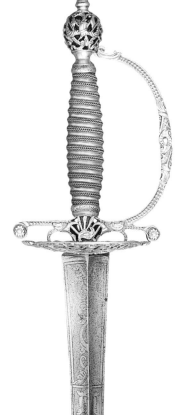

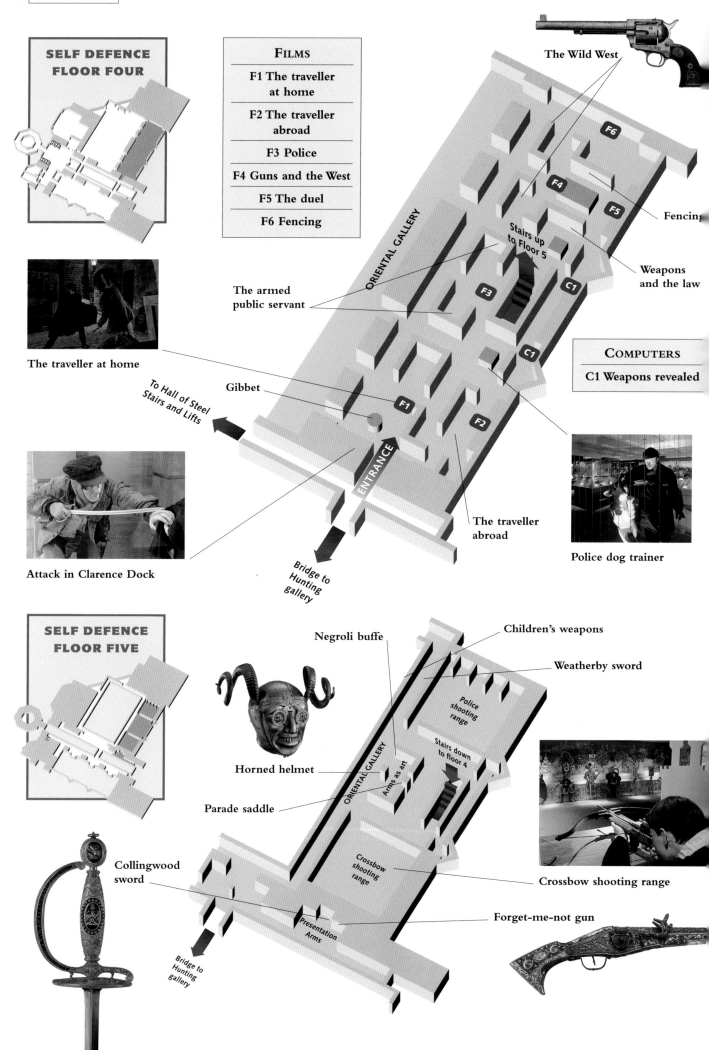

SELF DEFENCE FLOOR FOUR

FILMS

F1 The traveller at home

F2 The traveller abroad

F3 Police

F4 Guns and the West

F5 The duel

F6 Fencing

The Wild West

F6

F4

F5

Fencing

ORIENTAL GALLERY

Stairs up to Floor 5

Weapons and the law

C1

The armed public servant

F3

C1

COMPUTERS

C1 Weapons revealed

The traveller at home

To Hall of Steel Stairs and Lifts

Gibbet

F1

ENTRANCE

F2

The traveller abroad

Attack in Clarence Dock

Bridge to Hunting gallery

Police dog trainer

SELF DEFENCE FLOOR FIVE

Negroli buffe

Children's weapons

Weatherby sword

Horned helmet

Police shooting range

ORIENTAL GALLERY

Stairs down to floor 4

Arms as art

Parade saddle

Crossbow shooting range

Collingwood sword

Crossbow shooting range

Presentation Arms

Forget-me-not gun

Bridge to Hunting gallery

ARMS AND ARMOUR AS ART

The upstairs section of the Self-defence gallery is devoted partly to practical aspects of arms and armour, comprising two shooting galleries. In one of these laser pistols are used with police firearms training films, and the other in which you can try shooting moving targets with crossbows.

The main theme of the displays in this section is the art of the armourer. Arms and armour were from earliest times objects of great price and high status, and from medieval times in Europe the finest of them have always been decorated. The displays here present a cross section of superbly decorated arms and armour. The great period of armour decoration was the 16th century, when some of the finest artists including Hans Holbein in London and the Parisian goldsmith Etienne Delaune spent much of their time creating designs for armour. In some of the most extraordinary pieces decorative techniques such as embossing, etching and gilding were combined on pieces of armour which were intended for use as well as display. Swords with their chivalric and religious associations were particularly appropriate subjects for such fine decoration, and the swords with decorated hilts and fittings displayed here range from presentation pieces to fashion accessories. Then there are decorated firearms, from some early examples such as the 'Forget-me-not' gun, to the most recent guns decorated by artists working today. Finally it is easy to forget that the children of the nobility in medieval and later Europe were not only dressed as if they were adults, but provided with miniature arms and armour to complete the effect. A small display of boys' weapons completes the display in this area of the museum.

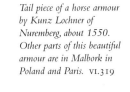

Tail piece of a horse armour by Kunz Lochner of Nuremberg, about 1550. Other parts of this beautiful armour are in Malbork in Poland and Paris. VI.319

Buffe by Filippo Negroli and brothers of Milan, signed and dated 1538. Embossed armour by this family of armourers represents the highest point of the artistic achievement of the armourer. IV.477

The 'Forget-me-not' wheellock gun, French, of about 1585. XII.1764

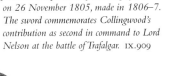

Left: *Presentation sword to Lord Collingwood from the Corporation of the City of London on 26 November 1805, made in 1806–7. The sword commemorates Collingwood's contribution as second in command to Lord Nelson at the battle of Trafalgar.* IX.909

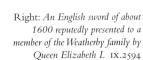

Right: *An English sword of about 1600 reputedly presented to a member of the Weatherby family by Queen Elizabeth I.* IX.2594

HUNTING

The human species began as hunter-gatherers, and ever since men and women around the world have hunted for subsistence, for profit or for sport. Hunting has had a profound effect upon landscapes and ecologies, and has both preserved and destroyed whole species of animals and birds. Hunting has bestowed on us a great tradition of legend and literature, and a rich heritage of music. It has also given us many of the breeds of horses and dogs which we own and love today. All types and conditions of mankind have hunted and do hunt.

There are many images of the hunter, from the self-reliant Leather Stocking living in harmony with nature in his backwoods to the callous big-game hunter on his relentless and insatiable pursuit of trophies. To many today hunting is repulsive and indefensible, to many it is right and natural, and to some it is still essential for survival. This is the story, both good and bad, which is told in this gallery.

About 10,000 years ago man first began to domesticate animals and plant crops. From settled farming communities producing an excess of food came the ability to specialise, to learn other craft skills, to live in towns, to become civilised. The change to settled agriculture and the development of civilisation led to increased populations which ever since have threatened animals by destroying their habitats.

Despite the change to a settled agricultural existence with its much reduced need to catch food to survive, man's love of hunting continued. It was in the blood.

'There is a passion for hunting, something deeply implanted in the human breast'

Charles Dickens

Above: Display of finely decorated hunting weapons.

Right: 'An Ugly Customer' – a recreation of a tiger hunt in India in the mid 19th century

HUNTING THROUGH THE AGES

Hunting evokes emotion at every level, and in this gallery items are displayed that will show the idealised aspects, such as 17th-century painting, 'The Prospect of Littlecote', which shows Littlecote House with various forms of hunting taking place in the fields, as well as the harsh reality of commercial and large-scale hunting, which can be seen in the display on whaling. Here you can see the weapons used in the hunting of whales, from early throwing spears to the harpoon cannons of today.

The history of hunting is traced from prehistoric times. A series of pictures from a book by Gaston Phoebus, written in 1387, are displayed, which describe medieval hunting techniques. This is some of the earliest evidence of how medieval hunts were conducted. The weapons associated with these early hunts can be seen; among them are a hunting crossbow, a boar sword and a hunting trousse used to prepare the dead game. Early decorated hunting guns, some of these rifled, can be seen here as well. On the other side of the gallery there are displays on boar hunting with spears, falconry and early target shooting.

Weapons for hunting were often given as expensive gifts, and some of the most finely decorated of the Museum's guns can be seen in the central area of the gallery. Especially important are the garniture made for Empress Elizabeth of Russia, a superb sporting gun by Simpson of York and a pair of late 17th-century pistols by the Huguenot gunmaker Pierre Monlong.

Above top: A German wheellock sporting gun of about 1620 by Nicolas Keucks. The stock is delicately inlaid with silver wire and stylised flowers. XII.1551

Above: An English flintlock gun made by William Simpson of York, about 1738. This is probably the finest provincially made sporting gun, with the use of chiselled rococo ornament and inlaid silver wire. XII.5843

One of a pair of English pistols by Henry Hadley, about 1755. These superb silver decorated pistols were commissioned by the Duke of Marlborough and show 18th-century English craftsmanship at its best. XII.1645–6

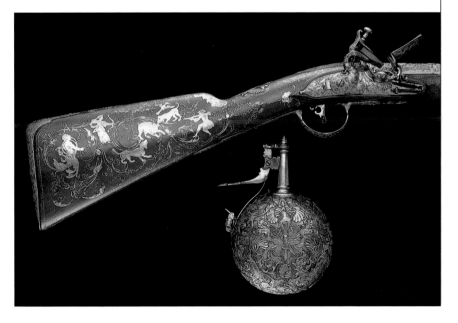

Part of a garniture made in Tula for the Empress Elizabeth of Russia and dated 1752. This set is decorated in the French style of chiselled steel on a gold background. XII.1504, XIII.150

HUNTING AS SPORT

Right: *This scene shows Walter Linnet, a professional wildfowler, hunting birds on the marshes of the Blackwater Estuary, Essex in the 1920s.*

Below right: *The charge, a print from Hog hunting in Lower Bengal by Percy Carpenter, 1861. This was a popular sporting pastime for the British in India.*

Below: *Hunting spear, French or Italian, about 1600. The lower part of the blade, the socket and the toggle are encrusted with gold and silver.* VII.81 *Next to this is a very ornate hunting hanger, probably Dutch and made between 1650 and 1670. The cast and chased silver hilt is in the form of a lion being attacked by hounds.* IX.849

The evolution of hunting into sport is a major theme, examining three types of hunting; the deer hunt, bird shooting and big-game hunting in Africa. There are films to explain how hunting can help to conserve environments both in Africa and Britain. Shooting as a competitive sport and the development of air weapons, examples of which range from the 18th century to the present day, are also covered in the displays. The gun room, situated at the end of the gallery, is designed to echo one from a late Victorian country house with a variety of weapons from many ages. The centrepiece of the gallery is a set of sporting guns from the armoury of the Dukes of Brunswick. There is also a fine display of crossbows including up-to-date models.

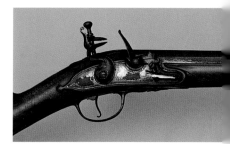

A German flintlock sporting gun made by Houle of Zella, about 1700. This is one of nine guns from the armoury of the Dukes of Brunswick and Luneburg. XII.6813

Several powerful and dramatic reconstructions demonstrate the wide diversity of hunting techniques and methods. On the main floor, the rhino hunt contrasts with the hunting of deer using a stalking horse. Two of the most impressive and evocative recreations are the tiger hunt on the main floor and the pig-sticking diorama on the upper level: here the almost obsessive interest in hunting as a pastime by the British in India is evoked. Also on the upper floor are the contrasting scenes of the Alpine chamois hunter and the lonely pursuit of the Essex wildfowler. The enormous size of the gun can be seen, as well as real examples in the accompanying case and a film on puntgunning.

HUNTING FLOOR FOUR

To Newsroom

Tiger Hunt

Stalking horse

Bridge to Oriental gallery

Stairs up to floor 5

Crossbows

To Hall of Steel, Stairs and Lifts

CAFE

ENTRANCE

Whaling gun

Rhino hunt

Bridge to Self-defence gallery

Medieval hunting

Cinema

F1

C1

F4

Dutch hunting hanger

F2

F2

C1

F3

Simpson of York gun

F3

Victorian gun room

COMPUTERS

C1 Tracker

Monlong pistols

Tula garniture

Stairs down to Floor 4

Lift to all floors

Bridge to Hunting and Self-defence galleries

HUNTING FLOOR FIVE

F5

F5

Pig-sticking

F6

F6

Punt gun

Chamois hunt

THE NEWSROOM

This multi-use facility on floor 4 specialises in the interpretation of current affairs. The versatile auditorium can be used for audio-visual displays and presentations, including film and video and broadcasts from around the world. Special events such as conferences, seminars and lectures are hosted here. Dramatic presentations relating to current themes and issues are also performed here providing a stimulating modern aspect to the live interpretation programme in the rest of the museum.

THE TILTYARD

Running alongside the River Aire for 150 metres, with seating on the landward side, is the Tiltyard. Here, weather permitting, we put on exhibitions of military and sporting skill at arms, including jousting. We also show the development of arms and armour through historical pageants, and demonstrate how animals – horses, dogs and hawks – have worked with man on the battlefield and in the hunting field.

Our intention in the Tiltyard is to give you an impression of how our collections were used in ways that would be impossible inside the museum. The performers you see mainly use replica weapons and armour which the museum is constantly acquiring. By trying to recreate the past we find out much more about it.

Above right: *Cavalry through the ages: an English Civil War trooper, a British Dragoon of 1815, a US cavalryman of 1876 and a British First World War trooper*

Above: *An interpreter in replica Elizabethan armour ready to test his skill.*

Right: *Lances 'shivering' during a 15th-century joust.*

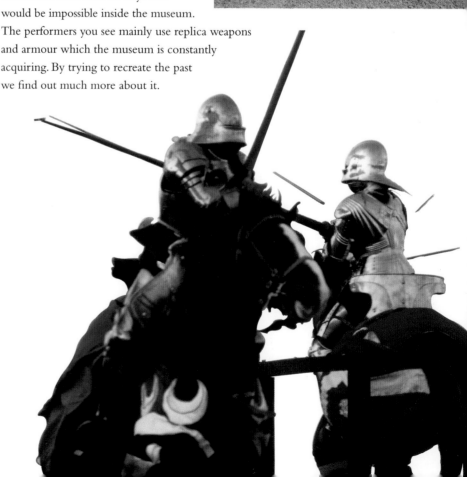

THE CRAFT COURT

The Royal Armouries Museum is more than just a collection in a building. Our objects were largely made for use out of doors, and therefore external demonstration areas have been provided to enable the collection to be properly put in context, explained and shown in action.

Next to the main building is the Craft Court where the visitors can see a selection of craftspeople working at their trades. These include: our resident gunmakers, who make and repair modern and antique guns, and provide many of the replicas our interpreters use in the museum; and a leather worker, making such things as footwear, scabbards and accessories used by soldiers and sportsmen through the ages, and the buff coats that were worn by so many soldiers for defence during the mid 17th century. There is also an armourer's workshop where the armours used in the museum's demonstrations are repaired and refurbished and where, from time to time, our own or visiting craftsmen show how metal is worked hot from a charcoal forge. All the craftspeople in the Craft Court have been carefully selected and are among the very best in the country. They work largely with modern tools, wearing modern dress, but using traditional techniques. You can watch them at work, but it is important to realise that most of them are here working for their living. They are not paid by the museum, but make their money by working at their craft.

The leather-worker in discussion with an interpreter, wearing one of the 17th-century-style buffcoats made in his workshop.

THE MENAGERIE COURT

Between the Tiltyard and the Craft Court is the Menagerie Court where the birds, dogs and horses that you see in the Tiltyard shows are housed. But here you can see them close at hand and meet and talk to their trainers and riders.

The Menagerie Court is named after the famous royal menagerie or zoo, which was housed in the Tower of London from the beginning of the 13th century until 1835.

The horses are our own, most specially selected to represent as accurately as we can the type of medieval war horse. You may be surprised at their size, but generally the medieval war horse was not particularly tall by present day standards. But they were like ours, strong, short backed and with thick, powerful necks. They have all been thoroughly trained to ensure that they are comfortable with swords and spears clashing next to their eyes and with the loud reports of firearms shot from just behind their heads.

We also own a small selection of hunting and gun dogs. However, the hawks and falcons you see belong to the falconer who works for us on contract.

All our animals are well cared for and properly trained. However they are animals and can be dangerous. Please take care of yourselves.

A falconer with saker falcon and Brittany spaniel.

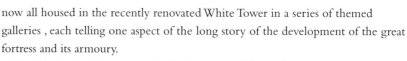
TOWER OF LONDON

The Royal Armouries has occupied buildings within the Tower of London for making and storing arms and armour for as long as the Tower itself has been in existence and for display purposes since at least the 15th century.

With the establishment of its other two museums in Leeds and Fort Nelson the Royal Armouries in the Tower now concentrates upon the display and interpretation of its historic Tower collections, its own long history in the Tower and the history of the Tower itself.

The public displays of the Royal Armouries in the Tower of London are now all housed in the recently renovated White Tower in a series of themed galleries , each telling one aspect of the long story of the development of the great fortress and its armoury.

- The Medieval gallery traces the development of the castle up to the 16th century.
- The Royal Armour gallery contains the best collection of royal armours in Britain.
- The Ordnance gallery tells the story of the Board of Ordnance, which supplied weapons to the British armed forces around the world from the 15th century until 1855.
- The Small Armoury recreates the mass displays of military weapons and equipment which decorated the Grand Storehouse.
- The Spanish Armoury recreates a propaganda display first established in the 17th century which purported to show weapons and instruments of torture taken from the Spanish armada of 1588; most of them in fact were taken from Henry VIII's arsenal in the Tower.
- The Line of Kings is an interpretation of the greatest of the Tower's propaganda displays of mounted royal armours.
- The Artillery Room displays cannon used or captured by British forces as they were displayed in the Grand Storehouse and tells the story of the fire which destroyed it in 1841.
- The Victorian Tower gallery contains several 19th-century models illustrating how the castle was changed and restored in the 19th century as it ceased to be a working arsenal.
- There is also a temporary exhibition gallery which houses a succession of Royal Armouries exhibitions.

Right: *The White Tower, the oldest part of the Tower of London, built about 1078–1100.*

Below: *The recently reconstructed Line of Kings display.*

Below right: *Armour of Henry VIII, made in the royal workshops at Greenwich, 1540.* II.8

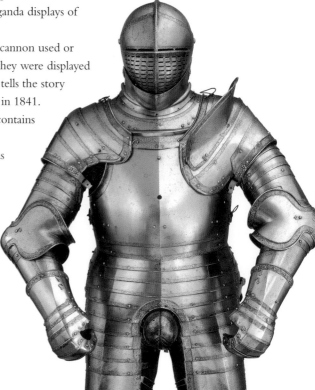

FORT NELSON

The Royal Armouries Museum at Fort Nelson was opened by His Grace the Duke of Wellington on 29 March 1995.

Fort Nelson, built in the 1860s to deter the French from attacking Portsmouth and its vital Royal Dockyard, is a scheduled ancient monument leased from Hampshire County Council. It has been carefully restored to house the Royal Armouries artillery collection. The artillery collection had its origins in the arsenal at the Tower of London. As the Tower gradually became a showplace obsolete artillery and trophies filled the fortress. Fort Nelson was chosen as an ideal site to display the collection properly, and with its vast Parade, to put on daily gun-firings and regular events through the summer months. Live interpretations take place daily.

Left: *The lower west gate of Fort Nelson, built by 1870.*

Impressive pieces of Victorian garrison armament have been mounted in original gun emplacements, fired by the Portsdown Artillery Volunteers on special event days.

Important 19th- and 20th-century guns have been acquired including sections of the imposing Iraqi 1000 mm Supergun on show with a medieval 'supergun' – the great Turkish bombard cast in 1464. These are shown in the Artillery Hall where the history of field guns is covered, including a 'Sexton' self-propelled gun of World War II. In the barrack block a series of galleries covers artillery from the earliest times to the end of the 19th century. Further displays show how guns were made and transported. Another gallery contains decorative guns, paintings and a tapestry depicting a gun foundry. Victorian garrison life is evoked by a restored barrack room, Officers' Mess kitchen, the elegant Officers' Mess itself and the main tunnel and underground magazines.

Above: *Members of the Portsdown Artillery Volunteers fire a British 16-pounder RML field gun of 1879 in the parade ground at Fort Nelson.* XIX.277

Left: *One of the many pieces from Ranjit Singh's impressive artillery train captured during the Sikh Wars. This field gun is on display complete with its original carriage and limber.* XIX.329

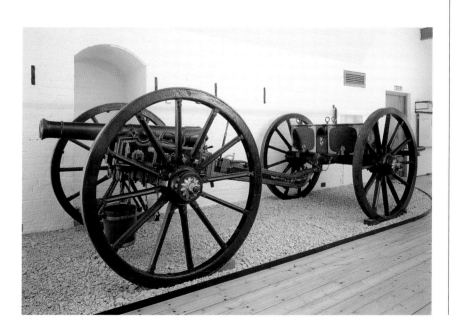

THE EDUCATION CENTRE

The Education Centre provides a wide range of services, catering for all ages from nursery to adult. Teacher's packs and a large number of worksheets and lesson styles offer support not just for those interested in history but also for students of art and design, technology and science – even maths! First-hand learning using both original and replica artefacts offers a stimulating and unique experience for all of our students. For further information telephone 0113 220 1888.

THE LIBRARY

The Royal Armouries library is a world-class facility comprising nearly 40,000 books, pamphlets and journals. Most of these relate directly to the Museum's collections, the history of arms and armour and the history of the Tower of London. The collection includes manuscripts and early books. The picture library mainly covers collection objects: there are about 150,000 black-and-white photographs and about 2,000 colour transparencies. The main library is housed in Leeds, and is open free of charge between 10.30 am and 4.30 pm, Monday to Friday (not including Bank Holidays). For further information please telephone 0113 220 1832.

For more information about the Royal Armouries visit our website at www.armouries.org.uk

Front cover: *Elephant armour, Indian, Mughal, about 1600.* XXVIA.102
Back cover: *The Royal Armouries Museum, Leeds.*

Designed by Royal Armouries Design
Royal Armouries Museum,
Armouries Drive, Leeds LS10 1LT
©2000 The Trustees of the Armouries
First published 2000, reprinted 2001
All rights reserved; no part of this publication may be reproduced, stored in any retrieval system, or transmitted in any form or by any means without the prior written permission of the Trustees of the Armouries.
ISBN 0 948092 43 2
Printed in Great Britain

A History of the Cathedral Church of All Saints Derby

Paul H. Bridges

Derby Cathedral
Enterprises Ltd